YOUR KNOWLEDGE HAS VALUE

- We will publish your bachelor's and
  master's thesis, essays and papers

- Your own eBook and book -
  sold worldwide in all relevant shops

- Earn money with each sale

Upload your text at www.GRIN.com
and publish for free

R. Kottaimuthu, N. Vasudevan

Endemic and Red-listed Medicinal Plants used by the Valaiyars of Karandamalai, Southern Eastern Ghats, Tamil Nadu, India

GRIN Verlag

Imprint:

Copyright © 2014 GRIN Verlag GmbH
Druck und Bindung: Books on Demand GmbH, Norderstedt Germany
ISBN: 978-3-656-60592-8

This book at GRIN:

http://www.grin.com/en/e-book/269170/endemic-and-red-listed-medicinal-plants-
used-by-the-valaiyars-of-karandamalai

ENDEMIC AND RED-LISTED MEDICINAL PLANTS USED BY THE VALAIYARS OF KARANDAMALAI, SOUTHERN EASTERN GHATS, TAMIL NADU, INDIA

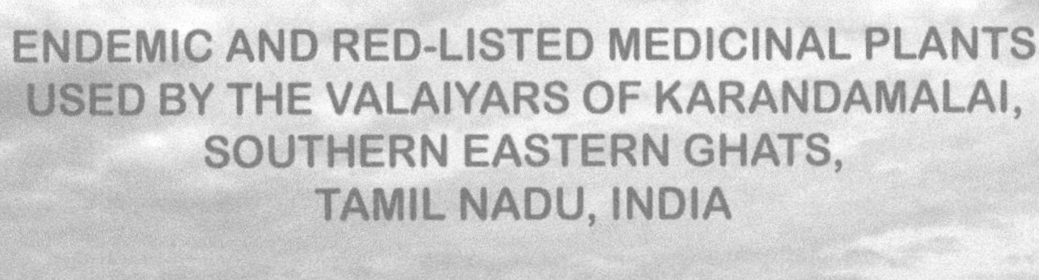

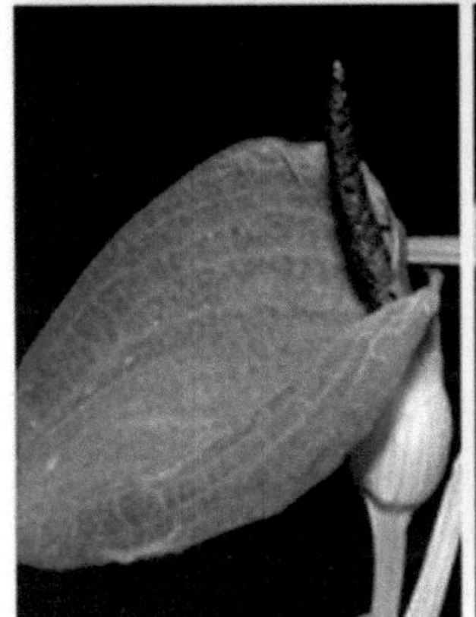

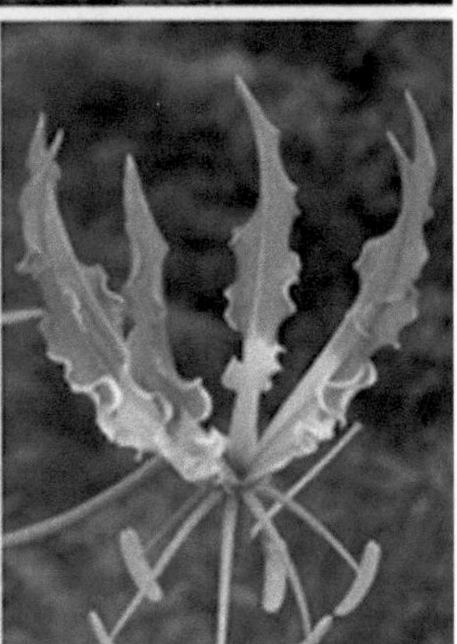

By
R. KOTTAIMUTHU
N. VASUDEVAN

Endemic and Red-listed Medicinal plants used by the Valaiyars of Karandamalai, Southern Eastern Ghats, Tamil Nadu, India.

R. Kottaimuthu* and N. Vasudevan

Ashoka Trust for Research in Ecology and the Environment (ATREE), Royal Enclave, Sriramapura, Jakkur, Bengaluru-560 064, Karnataka, India.
*Present Address: Centre for Research in Botany, Department of Botany, Saraswathi Narayanan College, Madurai-625 022, Tamil Nadu, India.

Abstract

The paper enumerates the traditional uses of endemic and red-listed medicinal plants utilized by the Valaiyars of Karandamalai for primary health care. Intensive ethnobotanical surveys through structured questionnaires were conducted to tap the ethnobotanical wisdom of Valaiyars. Information on the medicinal uses of 45 plants belonging to endemic and red-list category was documented. Vernacular names of the plants, parts used, methods of medicine preparation, dosage, mode and time of administration were also gathered. Ethnomedicinal plants are arranged alphabetically followed by botanical name, family name, local name and parts used, mode of preparation and medicinal uses.

Key words: Endemic, Red-list, Karandamalai, Medicinal plants, Valaiyars.

1. Introduction

India has an ancient, rich and diverse living tradition in the use of medicinal plants. The country has about 19,395 flowering plants[1], of these 5,725 species are endemic[2]. Nearly 75% of the drugs mentioned in various pharmacopoeias are growing in India. Medicinal plants continue to be an important source of life saving drugs for human kind, especially in the developing nations. The World Health Organization has estimated that more than 80% of the world population in developing countries depends on plants for basic health care[3]; 85% of the medicines in primary healthcare are derived from plants[4]. In rural India, 70% of the population is dependent on the traditional system of medicine[5]. The increasing realization of the health hazards and toxicity associated with the indiscriminate use of synthetic drugs and antibiotics has renewed the interest in the use of plants and plant-based drugs. Subsequent global inclination toward herbal medicine has advanced the expansion of plant-based pharmaceutical industries.

The obligatory demand for a huge raw material of medicinal plants is met from their wild populations. Over-exploitation and destructive harvesting to meet such demand in fact threaten the survival of many rare species. There is no reliable estimate for the number of medicinal plants that are globally threatened, however the reported species ranges from 4160-10 000[5,6]. In 2008, IUCN Red List shows that the number of threatened plant species is increasing gradually. The number of threatened plants is 8457, out of which 247 plants are found at different biodiversity hotspots in India. Our estimates of endangered species of flowering plants in India have unfortunately sharply risen from few hundred to a few thousand species over the past few years[7]. Confronted by such unprecedented genetic erosion and disappearance of species and ecosystems, conservation of natural resources assumes paramount urgency. To undertake and promote focussed conservation action, FRLHT has been engaged in systematic and rapid assessment of threatened medicinal plant taxa of India through the CAMP (Conservation Assessment and Management Plan) exercises. This methodology has been designed and developed by CBSG (Conservation and Breeding Specialist Group) of Species Survival Commission of IUCN for rapid assessment of Red List status of prioritized taxa. During 1993-2003, 10 such workshops have been conducted from different states of India and 304 taxa of medicinal plants have been assessed and assigned Red List status ranging from Near Threatened to Extinct. Four of these workshops have been conducted from 1993-1997 for the 3 southern Indian states viz., Kerala, Karnataka and Tamil

Nadu and results of these assessments enlist 110 taxa of medicinal plants for conservation focus[8].

2. Topography of the Study area

Karndamalai is a part of Southern Eastern Ghats and situated 43 km from Madurai. It is adjoined by Ariyalur hills in the west and towards the northwest and northeast it is surrounded by the Sirumalai and Perumalai hills, respectively. It lies between 10^0 15' to 10^0 21' north latitude and 78^0 9' to 78^0 15'east longitude. The altitude from foothill to the highest Jandamedu ranges from 180 to 916 M with undulating terrain. Forest types range from tropical thorn forests to mixed deciduous forests and moist deciduous riparian forests[9,10].

2.1. Geology and Soil

Geographically, Karandamalai is an archaen formation (i.e. made up of gneissic rocks). The gneissic rocks are referred to as charnockite and consist of mica, feldspar and quartz. Red soil is the predominant soil type found in Karandamalai. The surface soil in the scrub forests is usually of red soil mixed with pebbles. In the deciduous forests, soil is greyish-yellow lateritic clay. The riparian forests have loamy blackish soil rich in humus.

2.2. Climate and Rain fall

Except the riparian areas, majority of the parts in Karandamalai is hot and dry. The relatively cool season is December, January and part of February. During this period, there is heavy due formation at nights and mornings are foggy. The hottest months are April-May. The annual mean temperature of this are varies during summer and winter from 25^0C-30^0C and 17^0C-26^0C. The rainfall regime is a tropical dissymmetric type with the bulk of rain received during the retreating monsoon period (October-December) due to depressions and cyclones. Some rain is also received during the South-west monsoon.

3. Ethnography of Valaiyar

Valaya or Valaiyars are one of the most ancient castes in the country and they were often found in the villages on periphery of forests in the 18th and 19th century. They belong to denotified tribe of Southern India and traditionally they are rat trappers. In the census report of 1901, the Valaiyars are described as a 'shikari (hunting) caste' in Madurai and Tanjore. They are one of the oldest aboriginal groups inhabiting the hill tracks of southern Tamil

Nadu[11]. They are skilled hunters and forest product gatherers and their name is believed to be derived from the word 'valai' (or net), since this implement is constantly employed by them in the capturing of jungle game[12]. They are the repositories of the knowledge of herbal medicine and they utilize local herbs for different ailments after centuries of trail.

4. Methodology

Intensive interviews were carried out in the field with the Valaiyars, following standard methodology[13-16]. The gathered data was cross-verified by repeated queries with different local herbalists in different seasons in order to validate the information. The collected plants were identified taxonomically with the help of various floras[17,18]. The ethnomedicinal plants were preserved or pressed immediately. The herbarium specimens were prepared as per the standard specification and their identification was later confirmed by matching specimens with previously authenticated specimens available at Botanical Survey of India, Southern Circle, Coimbatore. All collections are deposited in Ashoka Trust for Research in Ecology and the Environment (ATREE) Herbarium, Bangalore.

ENUMERATION OF ETHNOMEDICINAL PLANTS

The enumeration follows alphabetical order of the binomials. Family name is given in uppercase in parentheses and the local name (Tamil name) is in italics within inverted commas, followed by collectors initials (RKM- R. Kottaimuthu) and collection number.

1, ***Acalypha alnifolia*** Klein ex Willd. (Euphorbiaceae); 'Chinnikeerai' (RKM-1099); Leaves paste is taken internally for dysentery.

2. ***Andrographis ovata*** (T. Anders. ex Bedd.) Benth. ex Clarke (Acanthaceae); *'Periyanangai'* (RKM-1050); Leaf paste is applied topically for poisonous bites. 20-30 ml of leaf juice is given for diarrhoea.

3. ***Aphanamixis polystachya*** (Wall.) Parker (Meliaceae); *'Vella kongu'* (RKM-1085); Oil extracted from the seed is used to treat skin diseases, especially eczema.

4. ***Artocarpus hirsutus*** Lam. (Moraceae); *'Kattupala'* (RKM-1096); Fruits used as an appetizer. In addition, the powdered seed is mixed with honey and used in the treatment of asthma.

5. ***Barleria courtallica*** Nees (Acanthaceae); *'Kaattukanagambaram'* (RKM-1058); Root juice is given for poisonous bites.

6. ***Barleria longiflora*** L. f. (Acanthaceae); 'Vellakuringi' (RKM-1009); Leaf infusion is given for cough. Root paste is applied topically for poisonous bites.

7. ***Caralluma pauciflora*** (Wight) N. E. Br. (Asclepiadaceae); 'Puliyanpirnadai' (RKM-85); Stem is ground with water and taken internally for stomach disorder.

8. ***Caralluma stalagmifera*** C. E. C. Fischer (Asclepiadaceae) *'Pulichai'* (RKM-2022); 2-3 teaspoons stem paste mixed with a glass of hot water, given orally for a week for stomach ulcer.

9. ***Cassia montana*** B. Heyne ex Roth (Caesalpiniaceae); *'Malaiaavaram'* (RKM-1057); Leaf paste is mixed with neem oil and applied topically for scabies. Stem bark decoction is given for stomach complaints.

10. ***Catunaregam brandisii*** (Gamble) R. Kottaimuthu (Rubiaceae); *'Malaikaarai'* (RKM-270); Diluted leaf juice is given for dysentery. Fruits are used as fish poisons.

11. ***Celastrus paniculatus*** Willd. (Celastraceae); *'Valuluvai'* (RKM-1140); A decoction of the bark is given orally on an empty stomach for a period of 7 days to women for the purposes of abortion.

12. ***Commiphora berryi*** (Wight & Arn.) Engler (Burseraceae); *'Mulkiluvai'* (RKM-444); Powdererd resin is mixed with hot water and taken internally for dysentery.

13. ***Commiphora pubescens*** (Wight & Arn.) Engler (Burseraceae); *'Malaikiluvai'* (RKM-111); Dried resin is mixed with hot milk or honey and taken internally to arrest diarrhoea.

14. ***Cyanotis tuberosa*** (Roxb.) Schultes & Schultes (Commelinaceae); *'Valukaikizhangu'* (RKM-108) Tuberous root is ground with water and into paste and taken internally for a month for diabetes.

15. ***Decalepis hamiltonii*** Wight & Arn. (Asclepiadaceae); *'Maahaali kizhanku'* (RKM-1071); Tuberous root paste is given for stomach disorders.

16. ***Deccania pubescens*** (Roth) Tirveng. (Rubiaceae); 'Peikaarai' (RKM-100); Leaf paste is applied topically for cuts and wounds.

17. ***Dicliptera cuneta*** Nees (Acanthaceae); 'Thelekadi poondu' (RKM-814); Leaves and roots are chewed for poisonous bites especially scorpion bites.

18. ***Endostemon viscosus*** (Roth) M. Ashby (Lamiaceae); 'Senthulasi' (RKM-1099); Leaf jucie is applied externally to repel ticks.

19. ***Ficus dalhousiae*** Miq. (Moraceae) *'Kalitchi'* (RKM-1120) Bark paste is applied externally to mend cracks in the feet. Stem bark decoction is used for taking bath to get relief from body pain.

20. ***Gardenia gummifera*** L. f. (Rubiaceae); *'Kumbil'* (RKM-1512); The mixture of resin and sugar in hot milk is used to arrest diarrhoea. Bark paste applied for skin diseases especially for scabies.

21. ***Gloriosa superba*** L. (Colchiaceae); *'Kanvalipoo'* (RKM-857); 10-20g of paste of tuberous roots is taken internally by women for abortion.

22. ***Gymnema elegans*** Wight & Arn. (Asclepiadaceae); *'Sirukurinjan'* (RKM-800); Handful of fresh leaves ground with water and the paste is taken internally for 30 days for diabetes.

23. ***Hemidesmus indicus*** (L.) R.Br. var. ***pubescens*** (Wight & Arn.) Hook.f. (Periplocaceae); *'Nannari'* (RKM-799); Decoction extracted from the roots is taken in an empty stomach for two months for diabetes. Root paste is taken internally for stomach disorder.

24. ***Henckelia incana*** (Vahl) Spreng. (Gesneriaceae); *'Kalthamarai'* (RKM-1253); Leaf is ground in water and the juice is taken orally to treat fever.

25. ***Hydnocarpus pentandra*** (Buch. - Ham.) Oken (Flacourtiaceae); *'Maravettai'* (RKM-1094); Oil extracted from the seed is applied externally for joint pains. Fruits used as fish poison.

26. ***Jasminum trichotomum*** Heyne ex Roth (Oleaceae); *'Kakka-minni'* (RKM-194); Leaf paste is applied topically to heal wounds.

27. ***Jatropha tanjorensis*** Ellis & Saroja (Euphorbiaceae); *'Katamanakku'* (RKM-1209); Latex applied topically to heal wounds. Latex applied on teeth and gums for toothache.

28. ***Jatropha villosa*** Wight (Euphorbiaceae); *'Thanakan'* (RKM-1207); Latex used to cure mouth ulcer.

29. ***Knoxia wightiana*** Wall. ex Wight & Arn. (Rubiaceae); *'Venthaadikulai'* (RKM-1066); A glass of leaf juice is given for dysentery. Root paste is taken internally in an empty stomach for a week for leucorrhoea.

30. ***Leucas hirta*** (Heyne ex Roth) Spreng. (Lamiaceae); *'Kaattuthumbai'* (RKM-1072); Leaf decoction is given to cure fever.

31. ***Manilkara roxburghiana*** (Wight) Dubard (Sapotaceae); *'Kanupaalai'* (RKM-1076); Latex applied on teeth and gums for toothache. Stem bark decoction is taken internally for diarrhoea.

32. ***Micrageria wightii*** Benth. (Scrophulariaceae); 'Vettukkaya thazhali' (RKM-1070); Leaf paste is applied externally for cuts and wounds. Root paste is taken internally twice a day for three days against dysentery.

33. ***Miliusa eriocarpa*** Dunn (Annonaceae); 'Sakkada maram' (RKM-19); Leaf paste is applied externally to get relief from joint pains.

34. ***Phyllanthus indofischeri*** Bennet (Phyllanthaceae); *'Nelli'* (RKM-1147); A handful of tender leaves ground with water and the paste is mixed with honey and given to children to arrest dysentery. 50 ml of fruit juice is taken internally for stomachache.

35. *Phyllanthus kozhikodianus* Sivar. & Manilal (Phyllanthaceae); 'Keezhanelli' (RKM-1109); Plants ground with turmeric and made into pills; 3-5 pills 3 times a day for a week to cure jaundice.

36. *Pseudarthria viscida* (L.) Wight & Arn. (Fabaceae); *'Kodiottai'* (RKM-1252); A handful of fresh leaves made in to a paste, mixed in a glass of hot water, orally administered twice in a day for stomach disorders. Leaf paste is mixed with rice gruel and taken internally by women for lactation.

37. *Radermachera xylocarpa* (Roxb.) K. Schum. (Bignoniaceae); (RKM-198); Stem bark decoction is taken internally to reduce the fever.

38. *Santalum album* L. (Santalaceae); *'Sandhanam'* (RKM-543); Stem paste is applied to purge pimples on the face.

39. *Schrebera swietenioides* Roxb. (Oleaceae); *'Pasaripattai'* (RKM-1040); Root paste about 10g is taken in empty stomach continuously 7 days for the treatment of Leucorrhoea.

40. *Strobilanthes ciliates* Nees. (Acanthaceae); *'Kurinji'* (RKM-870); Leaf paste mixed with neem oil is applied for eczema.

41. *Syzygium alternifolium* (Wight) Walp. (Myrtaceae); *'Kaattunavval'* (RKM-808); Seed powder is taken in an empty stomach for 45 days for diabetes. 10-15 fruits are consumed internally for one month to purify the blood.

42. *Tamilnadia uliginosa* (Retz.) Tirveng. & Sastry (Rubiaceae); 'Peikaarai' (RKM-504); Diluted leaf juice is given for dysentery. Fruits used as fish poison.

43. *Terminalia cuneata* Roxb. (Combretaceae); *'Maruthu'* (RKM-958); Paste made from an equal quantity of leaves and tender fruits mixed with honey is taken twice in a day for 30 days for curing diabetes. 20-30g of stem bark paste is taken internally for leucorrhoea.

44. *Theriophonum fischeri* Sivad. (Araceae); *'Kaattukarunai'* (RKM-601); Boiled root tubers are cooked and consumed for a period of 15days to cure piles.

45. *Tragia bicolor* Miq. (Euphorbiaceae); *'Senthatti'* (RKM-941); The juice of the root is taken orally to get relief from constipation. Dried root powder mixed with goat milk is administered to arrest frequent urination.

5. Results and Discussion

The present investigation reveals that the Valaiyars of Karandamalai use 45 endemic and red-listed medicinal plants for the treatment of various ailments. Of these, 32 species are endemic to Peninsular India; six are endemic and red-listed medicinal plants and seven non-endemic red-listed medicinal plants (Table 1). These plants belonged to 15 families. There were 5 species from Rubiaceae, 5 species from Acanthaceae, 4 species from Euphorbiaceae; 3 species from Asclepiadaceae; 2 species each from Lamiaceae, Moraceae, Oleaceae, Periplocaceae and Phyllanthaceae. The families Santalaceae, Flacourtiaceae, Gesneriaceae, Celastraceae, Colchiaceae and Caesalpiniaceae are represented by two species each. Rest of the families viz., Annonaceae, Araceae, Bignoniaceae, Caesalpiniaceae, Celstraceae, Colchiaceae, Combretaceae, Commelinaceae, Fabaceae, Flacourtiaceae, Gesneriaceae, Meliaceae, Myrtaceae, Santalaceae and Scrophulariaceae are represented by single species each. Among the plants utilized, 18 are trees, 12 herbs, 7 climbers, and 8 shrubs (Fig.1). Different parts of plants like leaves, flowers, fruits, bark, resin, latex etc. are being used for different purposes. Analysis of plant parts used by the Valaiyars shows that leaves and roots are the most widely used parts followed by fruits, bark, stem, resin and whole plant. The recorded plants are being administered for different health problems such as body pain, diabetes, tooth ache, gastro-intestinal problems (constipation, diarrhoea, dysentery, stomach disorder, stomach-ache and ulcer), gynaecological complaints (abortion, lactation and leucorrhoea), fish poisons, jaundice, joint pains, poisonous bites, respiratory problems (asthma, cough and fever), skin problems (cracks, cuts and wounds, eczema, pimples and scabies), ticks repellent and urinary complaints. Seven species of plants are used to treat dysentery, 5 for diabetes, 4 for cuts and wounds, 3 for fever, poisonous bites and stomach disorder, 2 for abortion, diarrhoea, eczema, joint pains and scabies and only 1 for the treatment of the following sicknesses or disorders: asthma, body pain, constipation, cough, cracks, jaundice, lactation, leucorrhoea, piles, tooth ache and urinary troubles (Table 2). The information collected for some plants from the study area corroborate with the previous reports[19-27]. However, the medicinal uses of *Commiphora berryi* (Wight & Arn.) Engler, *Commiphora pubescens* (Wight & Arn.) Engler, *Ficus dalhousiae* Miq., *Gymnema elegans* Wight & Arn., *Manilkara roxburghiana* (Wight) Dubard and *Strobilanthes ciliatus* Nees are being reported for the first time in this report as there is no such record in literature[28-68].

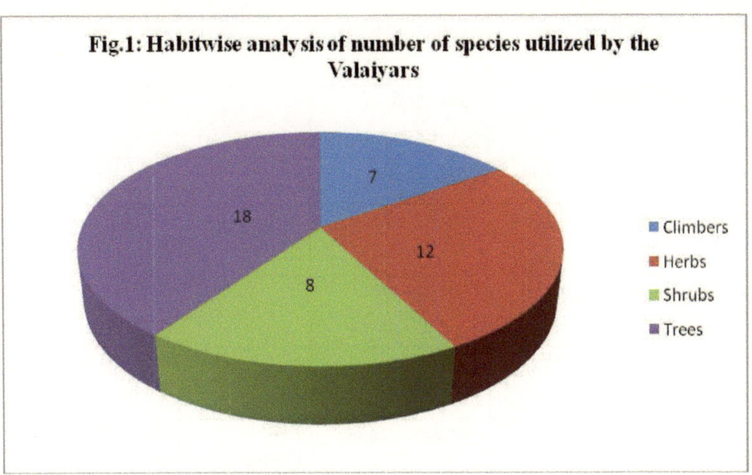

Table 1. List of Endemic and Red-listed Medicinal Plants and their utilization by Valaiyars of Karandamalai, Souther Eastern Ghats.

Botanical Name	Family	Endemic / Red List Category	Forest type
Acalypha alnifolia Klein ex Willd.	Euphorbiaceae	E	SC
Andrographis ovata (T. Anders. ex Bedd.) Benth. ex Clarke	Acanthaceae	E	SC
Aphanamixis polystachya (Wall.) Parker	Meliaceae	DD	MD
**Artocarpus hirsutus* Lam.	Moraceae	VU-Globally	MD
Barleria courtallica Nees	Acanthaceae	E	MD
Barleria longiflora L. f.	Acanthaceae	E	SC
Caralluma pauciflora (Wight) N. E. Brown	Asclepiadaceae	E	SC
Caralluma stalagmifera C. E. C. Fischer	Asclepiadaceae	E	SC

Cassia montana B. Heyne ex Roth	Caesalpiniaceae	E	SC
Catunaregam brandisii (Gamble) R. Kottaimuthu	Rubiaceae	E	DD
Celastrus paniculatus Willd.	Celastraceae	NT	SC; DD
Commiphora berryi (Wight & Arn.) Engler	Burseraceae	E	SC
Commiphora pubescens (Wight & Arn.) Engler	Burseraceae	E	SC; DD
Cyanotis tuberosa (Roxb.) Schultes & Schultes	Commelinaceae	E	SC; DD
**Decalepis hamiltonii* Wight & Arn.	Periplocaceae	EN-Globally	DD
Deccania pubescens (Roth) Tirveng.	Rubiaceae	E	DD
Dicliptera cuneata Nees	Acanthaceae	E	SC
Endostemon viscosus (Roth) Ashby	Lamiaceae	E	SC
Ficus dalhousiae Miq.	Moraceae	E	MD
**Gardenia gummifera* L. f.	Rubiaceae	VU-Globally	DD
Gloriosa superba L.	Colchiaceae	LC	SC
Gymnema elegans Wight & Arn.	Asclepiadaceae	E	DD; DEG
Hemidesmus pubescens Wight & Arn.	Periplocaceae	E	SC; DD
Henckelia incana (Vahl) Spreng.	Gesneriaceae	E	SC; DD
**Hydnocarpus pentandra* (Buch. - Ham.) Oken	Flacourtiaceae	VU-Globally	MD
Jasminum trichotomum Heyne ex Roth	Oleaceae	E	SC
Jatropha tanjorensis Ellis & Saroja	Euphorbiaceae	E	SC

Jatropha villosa Wight	Euphorbiaceae	E	SC
Knoxia wightiana Wall. ex Wight & Arn.	Rubiaceae	E	DD
Leucas hirta (Heyne ex Roth) Spreng.	Lamiaceae	E	DD
Manilkara roxburghiana (Wight) Dubard	Sapotaceae	E	DD
Micragereia wightii Benth.	Scrophulariaceae	E	SC
Miliusa eriocarpa Dunn	Annonaceae	E	MD
Phyllanthus indofischeri Bennet	Phyllanthaceae	E	DD
Pseudarthria viscida (L.) Wight & Arn.	Fabaceae	NT	SC; DD
Radermachera xylocarpa (Roxb.) K. Shum.	Bignoniaceae	E	DD
Santalum album L.	Santalaceae	EN	DD
Schrebera swietenioides Roxb.	Oleaceae	DD	DD
Strobilanthes ciliates Nees	Acanthaceae	EN-Globally	MD
Syzygium alternifolium (Wight) Walp.	Myrtaceae	E	DD
Tamilnadia uliginosa (Retz.) Tirveng. & Sastry	Rubiaceae	E	DD
Terminalia cuneata Roxb.	Combretaceae	LC	MD
Theriophonum fischeri Sivad.	Araceae	E	SC
**Tragia bicolor* Miq.	Euphorbiaceae	EN	MD

E-Endemic; ** = Endemic and Red-listed medicinal plant; # = Non-endemic and red-listed medicinal plant; DD=Data deficient; VU-Vulnerable; EN=Endangered; LC-Least concern; NT=Near Threatened.

Table 2. Ailment category and herbal remedies of Valaiyars

Sl. No	Ailment category	Total number of Herbal remedies
1	Abortion	2
2	Appetizer	1
3	Asthma	1
4	Blood purifier	1
5	Body pain	1
6	Constipation	1
7	Cough	1
8	Cracks	1
9	Cuts and wounds	4
10	Diabetes	5
11	Diarrhoea	2
12	Dysentery	7
13	Eczema	2
14	Fever	3
15	Fish poison	3
16	Jaundice	1
17	Joint pains	2
18	Lactation	1
19	Leucorrhoea	1
20	Mouth ulcer	1
21	Piles	1
22	Pimples elimination	1
23	Poisonous bites	3
24	Scabies	2
25	Stomach-ache	1

26	Stomach disorder	3
27	Stomach ulcer	1
28	Ticks repellent	1
29	Tooth-ache	1
30	Urinary complaints	1

6. Conclusion

Many of today's modern medicines are derived from plants. In these days, great emphasis is being laid on documentation of the traditional knowledge of ethnic people and adoption of that knowledge in bio prospecting of biological resources as a new source of food and medicine. Pharmaceutical research seeks to profit from the vast knowledge of plants accumulated by natural healers. The present study assumes greater importance in enhancing our knowledge about the plants used by the ethnic community Valaiyars of Karandamalai and the threatened status of ethno medicinal plants for their conservation. The medicinal plants and their utilization in local health traditions by Valaiyars are gradually dwindling due to modernization and the diminishing interest of younger generations. Hence, there is an imperative necessity to conserve the rare and lesser known medicinal plants by measures safeguarding, documenting and preserving the traditional knowledge. Further, research on ethnomedicinal plants for their active principles, pharmacology and evaluation of efficacy would help in developing new plant drugs for posterity.

PLATE-1
A-*Acalypha alnifolia* Klein ex Willd.
B-*Artocarpus hirsutus* Lam.
C-*Barleria courtallica* Nees; D-*Barleria longiflora* L. f.
E-*Caralluma pauciflora* (Wight) N. E. Brown;
F-*Caralluma stalagmifera* Fisch.
G-*Commiphora berryi* (Wight & Arn). Engler
H-*Commiphora pubescens* (Wight & Arn.) Engler
I-*Catunaregam brandisii* (Gamble) R. Kottaimuthu

PLATE-2
A-*Cassia montana* Heyne ex Roth; B-*Gloriosa superba* L.
C-*Gymnema elegans* Wight & Arn.;
D-*Cyanotis tuberosa* (Roxb.) Schultes ex Schultes
E-*Deccania pubescens* (Roth) Tirveng.;
F-*Endostemon viscosus* (Roth) M. Ashby;
G-*Ficus dalhousiae* Miq.; H-*Gardenia gummifera* L. f.
I-*Henckelia incana* (Vahl) Spreng.

Acknowledgement

The first author is grateful to the Valaiyars of Karandamalai for their valuable help in documentation of indigenous ethnomedicinal knowledge and Dr. R. Ganesan, Scientist, Ashoka Trust for Research in Ecology and the Environment (ATREE), Bangalore for facilities and encouragement.

PLATE-3
A-*Hemidesmus indicus* var. *pubescens* (Wight & Arn.) Hook. f.
B-*Syzygium alternifolium* (Wight) Walp; C-*Terminalia cuneata* Roxb.
D-*Jasminum trichotomum* Heyen ex Roth.
E-*Jatropha tanjorensis* Ellis & Saroja.
F-*Leucas hirta* (Heyne ex Roth) Spreng.
G-*Manilkara roxburghiana* (Wight) Dubard; H-*Santalum album* L.
I-*Tragia bicolor* Miq.

Bibliography

[1] Karthikeyan S, Sanjappa M, Moorthy S. *Flowering plants of India, Dicotyledons. Vol. 1. (Acanthaceae-Avicenniaceae).* Botanicaly Survey of India, Kolkatta; 2009.

[2] Nayar MP. *Hot spots of Endemic plants of India, Nepal & Bhutan.* TBGRI, Thiruvananthapuram; 1996.

[3] Balick MJ, Cox PA. Plants, people and culture: the science of ethnobotany. Freeman and Co., Scientific American Library, New York;1996.

[4] Balick MJ, Korenberg F, Osoki, AL, Reiff M, Berman AF, Connor B, Roble M, Lohr P, Atha D. Medicinal plants used by Latino Healers for woman's health conditions in New York city. *Econ. Bot.* 2000; 54: 344-357.

[5] Ali M. Present status of herbal medicines in India. *J. Herbal Med. Toxic.* 2007; 3(2): 1-7.

[6]Natesh S. Conservation of medicinal and aromatic plants in India- an overview. In: Kamaruddin MS. *et al.* (eds) *Medicinal and aromatic plants: strategies and technologies for conservation.* Forest Research Institute, Kuala Lumpur; 1999.

[7]Gopalan R, Henry AN. *Endemic Plants of India: Camp for the strict endemics of Agasthiyamalai Hills, SW Ghats.* Bishen Singh, Mahendra Pal Singh, Dehra Dun; 2000.

[8]Ravikumar K, Ved DK. *100 Red-Listed Medicinal Plants of Conservation Concern in Southern India.* FRLHT, Bangalore; 2000.

[9] Champion HG, Seth SK. *A Revised Survey of the Forest Types of India.* Reprinted Edition, Natraj Publishers, Dehra Dun; 2005.

[10] Kottaimuthu R. *Systematic studies on the Dicotyledonous Flora of Karandamalai, Southern Eastern Ghats, Tamil Nadu.* M. Phil., Thesis, Periyar University, Salem; 2007.

[11] Thurston E, Rangachari K. *Castes and Tribes of Southern India.* Volume- VII, The Superindent, Government press, Madras; 1909.

[12] Kottaimuthu R. Ethnobotany of the Valaiyans of Karandamalai, Dindigul District, Tamil Nadu, India. *Ethnobot. Leaflets* 2008;12: 195-203.

[13] Schultes RE. Tapping our heritage of ethnobotanical lore. *Econ. Bot.* 1960; 14: 257-262.

[14] Schultes RE. The lore of the Ethnobotanist in search for new medicinal plants. *Lloydia* 1962; 25: 257- 366.

[15] Jain SK. Studies in Indian Ethnobotany plants used in by the tribals of Madhya Pradesh. *Bull. Regional Research Lab.* 1963; 1: 126-129.

[16] Jain SK. Methods and Approaches in Ethnobotany. Society of Ethnobotanists, CDRI, Lucknow; 1989.

[17] Gamble JS. *Flora of the Presidency of Madras.* Reprinted Edition, Volumes I-III, Bishen Singh Mahendra Pal Singh, Dehra Dun; 1997.

[18] Matthew KM. *An Excursion Flora of Central Tamil Nadu, India.* Oxford and IBH Publishing co., New Delhi; 1991.

[19] Balasubramanian P. Observations on the utilization of Forest plants by the tribals of Point Calimere Wild life Sanctuary, Tamil Nadu. *Bull. Bot. Surv. India* 1992; 34: 100-111.

[20] Ganesan S, Kesavan L. Ethnomedicinal plants used by the ethnic group Valaiyans of Vellimalai hills (Reserved Forest), Tamil Nadu, India. *J. Econ. & Taxon. Bot.* 2003; 27(3): 754-760.

[21] Ganesan S, Ramar Pandi N, Banumathi N. Ethnomedicinal Survey of Alagarkoil Hills (Reserved Forest), Tamil Nadu, India. *Electonic Journal of Indian Medicine* 2007; 1: 1-19.

[22] Kottaimuthu R. Suresh K. Ethnobotanical study of Medicinal Plants used by Valayar tribals of Alagar hills, Tamil Nadu, India. *Plant Archives* 2009; 9(2): 669-671.

[23] Rajendran SM, Chandrasekar M, Sundresan V. Ethnomedicinal lore of Valaya tribe in Seithur hills of Virudhunagar district, Tamil Nadu, India. *Ind. J. Trad. Know.* 2002; 1(1): 59-71.

[24] Sandhya B, Thomas S, Isabel W, Shenbagarathai R. Ethnomedicinal plants used by the Valaiyan community of Piranmalai Hills (Reserved Forest), Tamil Nadu, India.- A pilot study. *Afr. J. Trad. CAM* 2006; 3(1): 101-114.

[25] Subramanian A. Herbal medicinal plants used by the Valaiyans of Madurai district, Tamil Nadu to obtain relief from the poisonous bites *J. Econ. Taxon. Bot.* 2005; 29(2): 419-421.

[26] Subramanian A, Mohan VR, Kumaresan S, Chelladurai V. Medicinal plants used by the Valaiyans of Madurai District, Tamil Nadu. *J. Econ. Tax. Bot.* 2003; 27(3): 785-787.

[27] Subramaniam A, Mohan VR, Kalidass C, Maruthupandian A. 2010. Ethnomedico botany of the Valaiyans of Madurai district, Tamil Nadu. *Journal Econ. Taxon. Bot.* 34 (2): 363 – 379.

[28] Henry AN, Hosagoudar VB, Ravikumar K. Ethno-Medico-Botnay of the Southern Western Ghats of India. In: Jain SK. (ed.). *Ethnobiology in Human Welfare.* Deep Publications, New Delhi; 1996.

[29] Rajan S, Jayendran M, Sethuraman M. Medico-ethnobotany: A study on the *Kattunayaka* tribe of Nilgiri Hills, Tamil Nadu. *Nat. Remedies* 2003; 3(1): 68-72.

[30] Ganesan S, Venkatesan G, Banumathy N. Medicinal plants used by ethnic group *Thottinaickans* of Semmalai hills (Reserved forest), Tiruchirapalli District, Tamil Nadu. *Ind. J. Trad. Know.* 2006; 5(2): 245-252.

[31] Udayan PS, Tushar KV, George S, Balachandran I. Ethnomedicinal information from *Kattunayakas* tribes of Mudumalai Wildlife Sanctuary, Nigiri District, Tamil Nadu. *Ind. J. Trad. Know.* 2007; 6(4): 574-578.

[32] Udayan PS, George S, Thushar KV, Indiara Balachandar B. Ethnomedicine of Chellipale community of Namakkal district, Tamil Nadu. *Ind. J. Trad. Know.* 2005; 3(3): 437-442.

[33] Udayan PS, George S, Thushar KV, Indiara Balachandar B. Medicinal plants used by the Malayali tribe of Servarayan hills, Yercaud, Slaem District, Tamil Nadu, India. *Zoos'print Journal* 2006; 21(4): 2223-2224.

[34] Kathirvel K, Ramya S, Sudha TPS, Ravi AV, Rajasekaran C, Vanitha Selvai R, Jayakumararaj R. Ethnomedicinal survey on plants used by the tribals in Chitteri Hills. *Environ. We Int. J. Sci. Tech.* 2010; 5: 35-46.

[35] Ravikumar K, Sankar RV. Ethnobotany of Malayali tribes in Melpattu village, Javvadhu hills of Eastern Ghats, Tiruvannamalai district, Tamil Nadu. *J. Econ. Taxon. Bot.* 2003; 27(3): 715-726.

[36] Ramya S, Rajasekaran C, Sivaperumal R, Krishnan A, Jayakumararaj R. Ethnomedicinal perspectives of botanicals used by Malayali tribes in Vattal hills of Dharmapuri (Tamil Nadu), India. *Ethnobot. Leaflets* 2008; 12: 1054-1060.

[37] Suresh K, Norman TSJ, Velumani K, Vijayan R, Rathinavel S. Ethnobotany and livelihood status of Malayali tribes of Kollihills in Tamil Nadu. *Plant Archives* 2008; 8(1): 479-481.

[38] Suresh K, Norman TSJ, Velumani K, Vijayan R, Rathinavel S. Ethnomedicinal practices of Malayali tribes of Yercaud hills in Tamil Nadu. *Plant Archives* 2008; 8(1): 457-459.

[39] Dwarakan P, Ansari AA. Less known uses of Plants of Kollimalai. *J. Econ. Taxon. Bot.* 1996; 12: 284-286.

[40] Viswanathan MB. Ethnobotany of the Malayalis in the Yelagiri hills of North Arcot district, Tamil Nadu. *J. Econ. Taxon. Bot.* 1989; 13: 667-671.

[41] Viswanathan MB. Ethnobotany of the Malayalis in North Arcot district, Tamil Nadu, India. *Ethnobotany* 1997; 9: 77-79.

[42] Anand RM, Nandakumar N, Karunakaran L, Ragunathan M, Murugan V. A survey of medicinal plants in Kollimalai hill tracts, Tamil Nadu. *Natural Product Radiance* 2006; 5(2): 139-143.

[43] Ignachimuthu S, Ayyanar M, Sivaraman KS. Ethnobotanical investigations among tribes in Madurai District of Tamil Nadu (India). *J. Ethnobio. Ethnomed.* 2006; 2: 25 doi:10.1186/1746-4229-2-25.

[44] Ayyanar M, Ignachimuthu S. Traditional knowledge of Kani tribals in Kouthalai of Tirunelveli hills, Tamil Nadu, India. *J. Ethnopharmacology* 2005; 102(2): 246-255.

[45] Ignachimuthu S, Sivaraman KS, Kesavan L. Medico-ethnobotanical survey among Kanikar tribals of Mundanthurai Sanctuary. *Fitoterapia* 1998; 69: 409-414.

[46] Rajan S, Sethuraman M, Mukherjee PK. Ethnobiology of the Nilgiri hills, India. *Phytotherapy Research* 2002; 16(2): 98-116.

[47] Ganesan S, Suresh N, Kesavan L. Ethnomedicinal survey of lower Palni Hills of Tamil Nadu. *Ind. J. Trad. Know.* 2004; 3(3): 299-304.

[48] Ayyanar M, Ignachimuthu S. Ethnomedicinal plants used by the tribals of Tirunelveli hills to treat poisonous bites and skin diseases. *Ind. J. Trad. Know.* 2005; 2(4): 229-236.

[49] Muthukumarsamy S. Mohan VR, Kumaresan S, Chelladurai V. Herbal medicinal plants used by Paliyars to obtain relief from gastro-intestinal complaints. *J. Econ. Taxon. Bot.* 2003; 27(3): 711-714.

[50] Karuppusamy S. Medicinal plants used by Paliyan tribes of Sirumalai hills of Southern India. *Natural Product Radiance* 2007; 6(5): 436-442.

[51] Karuppusamy S, Kumuthakalavalli R. Some folklore medicinal claim on rare plants of Sirumalai hills, Tamil Nadu, India. *Bull. Med.-Ethno Bot. Res.* 1998; 3-4: 145-150.

[52] Jeyaprakash K, Ayyanar M, Geetha KN, Sekar T. Traditional uses of medicinal plants among the tribal people in Theni District (Western Ghats), Southern India. *Asian Pacific Journal of Tropical Biomedicine* 2011; s20-25.

[53] Ignachimuthu S, Ayyanar M, Sivaraman KS. Ethnobotanical study of medicinal plants used by the Paliyar tribals in Theni District of Tamil Nadu, India. *Fitoterapia* 2008; 79: 562-568.

[54] Alagesaboopathi C. Ethnomedicinal plants and their utilization by villagers in Kumaragiri hills of Slaem District of Tamil Nadu, India. *Afr. J. Trad. CAM* 2009; 6(3): 222-227.

[55] Alagesaboopathi C. Ethnobotanical studies on useful plants of Sirumalai Hills of Eastern Ghats, Dindigul District of Tamil Nadu, Southern India. *Int. J. Biosci.* 2012; 2(2): 77-84.

[56] Maruthupandian A, Mohan VR. Observations of ethnomedicinal plants from Sirumalai hills in Western Ghats of Tamil Nadu, India. *J. Herb Med. Toxic.* 2010; 4(2): 89-92.

[57] Pandikumar P, Ayyanar M, Ignachimuthu S. 2007. Medicinal plants used by Malasar tribes of Coimbatore District, Tamil Nadu. *Ind. J. Trad Know.* 6(3): 575-582.

[58] Balasubramanian P, Rajasekaran A, Prasad SN. Folk medicine of the Irular of the Coimbatore forests. *Ancient Sci. Life* 1997; 16(3): 222-226.

[59] Senthilkumar M, Gurumoorthi P, Janardhanan K. Some medicinal plants used by Irular, the tribal people of Maruthamalai hills, Coimbatore, Tamil Nadu. *Nat Prod Rad* 2006; 5(5): 382-388.

[60] Geetha S, Poornima S, Vaseegari J. Studies on the ethnobotany of Irulars of Anaikatty hills, Coimbatore District. *College Sci India* 2007; 1: 2-20.

[61] Rajendran A, Henry AN. Plants used by the tribe Kadar in Anamalai hills of Tamil Nadu. *Ethnobotany* 1994; 6: 19-24.

[62] Umapriya T, Rajendran A, Aravindhan V, Thomas B, Maharajan M. Ethnobotany of Irular tribe in Palamalai Hills, Coimbatore, Tamil Nadu. *Ind. J. Nat Prod Resour.* 2011; 2(2): 250-255.

[63] Shanmugam S, Ramar S, Ragavendhar K, Ramanathan R, Rajendran K. Plants used as medicine by Paliyar tribes of Shenbagathope in Virudhunagar district of Tamil Nadu. *J. Econ. Taxon. Bot.* 2008; 32(4): 922-929.

[64] Arunachalam G, Karunanithi M, Subramanian N, Ravichandran V, Selvamuthukumar S. Ethnomedicines of Kolli hills at Namakkal district in Tamil Nadu and its significance in Indian systems of medicine. *J. Pharm Sci & Res.* 2009; 1(1): 1-15.

[65] Karuppusamy S, Muthuraja G, Rajasekaran KM. Lesser known ethnomedicinal plants of Alagar hills, Madurai district of Tamil Nadu, India. *Ethnobot. Leaflets* 2009; 13: 1426-1433.

[66] Samuel JK, Andrews B. Traditional medicinal plant wealth of Pachalur and Periyur hamlets Dindigul district, Tamil Nadu. *Ind. J. Trad. Know.* 2010; 9(2): 264-270.

[67] Rao DM, Pullaiah T. Ethnobotanical studies on some rare and endemic floristic elements of Eastern Ghats hill ranges of South East Asia. *Ethnobot. Leaflets* 2007; 11: 52-70.

[68] Kottaimuthu R, Kumuthakalavalli R, Manikandan N. Ethnobotanical observations on Red-listed Medicinal Plants of Sirumalai Hills, Southern Eastern Ghats. In: Thilagavathy Daniel (Ed.), *Biodiversity: Richness, Uses, Threats and Conservation.* Excel India Publishers, New Delhi; 2012.